Ítalo Armando Pilay Ponce

Sustainable rural tourism model

Ítalo Armando Pilay Ponce

Sustainable rural tourism model

As a tourism initiative of the Puerto Cayo parish of the Jipijapa canton, province of Manabí-Ecuador

ScienciaScripts

Imprint

Any brand names and product names mentioned in this book are subject to trademark, brand or patent protection and are trademarks or registered trademarks of their respective holders. The use of brand names, product names, common names, trade names, product descriptions etc. even without a particular marking in this work is in no way to be construed to mean that such names may be regarded as unrestricted in respect of trademark and brand protection legislation and could thus be used by anyone.

Cover image: www.ingimage.com

This book is a translation from the original published under ISBN 978-613-9-41074-3.

Publisher:
Sciencia Scripts
is a trademark of
Dodo Books Indian Ocean Ltd. and OmniScriptum S.R.L publishing group

120 High Road, East Finchley, London, N2 9ED, United Kingdom
Str. Armeneasca 28/1, office 1, Chisinau MD-2012, Republic of Moldova, Europe
Printed at: see last page
ISBN: 978-620-7-95820-7

SUSTAINABLE RURAL TOURISM MODEL AS A TOURISM INITIATIVE IN THE PARISH OF PUERTO CAYO, CANTON JIPIJAPA, PROVINCE OF MANABÍ-ECUADOR

PRESENTATION

Sustainable rural tourism is one of the most demanded modalities by the tourist nowadays, due to the activities that are carried out in it and, above all, the experiences that are obtained through its realisation. In the case study, the rural parish of Puerto Cayo has several tourist attractions that can be sustainably exploited through this modality.

That is why this research proposal aims to promote sustainable rural tourism as a complementary tourism modality applicable in the locality, thus obtaining an optimal sustainable use of the tourism potential of the parish. In the same way, it seeks to promote the development of tourism in the area through the creation of products, activities and other proposals for sustainable rural tourism that are framed in a structured way in the action plan presented. Therefore, this will lead to the creation of employment, improving the economic situation, preserving resources and promoting the care of the natural environment.

Finally, it is important to mention that the implementation of this type of tourism will increase the level of investment in the parish, generating new sources of employment and a wide variety of products and services to offer. In addition, this will help to reduce the seasonality of demand in the rural parish of Puerto Cayo.

THANK YOU

I am grateful above all things to our creator and architect of the universe, God, for having endowed me with health, wisdom and intelligence to be able to carry out this research, which I have done with great pleasure for the good of the community in general. To the authorities of the Universidad Estatal del Sur de Manabí, Faculty of Economic Sciences and therefore to the Tourism Department for having allowed me to become a teacher. My special thanks to the community of the rural parish of Puerto Cayo, as well as to Marcos José Gutiérrez Bravo for having contributed to its development. The unconditional support I received from my family at all times is incalculable. To them my undying gratitude

Ítalo Armando Pilay Ponce

INTRODUCTION

Sustainable rural tourism is a fundamental element in local tourism development, positioning itself as a real alternative to the well-known and characteristic sun and beach tourism. The circumstances that have allowed this change in trend may be due to an infinite number of factors, whether personal, environmental, social, economic or geographical (Gutiérrez, 2021). In addition, rural tourism seen as sustainable tourism has several objectives, among which the following stand out: To make the population aware of the opportunities of rural tourism for the economy and the environment; to work for the protection of the environment, biotic and cultural; to contribute to the improvement of the standard of living of the community. According to Crosby (2009), "Rural tourism has the capacity to offer a luxury, which is increasingly necessary and will be more in demand by the rest of the urban population to enjoy the rural and natural environment, when it has not been lost or transformed" (p.15).

For Crosby (2009), "rural tourism grew in the 20th century as a reaction to increasing urbanisation and industrial activity. Poets and artists began to revalue rural life and landscapes" (p. 23). Thomé (2008), states that rural tourism activity coexists with multiple realities that occur in the same space, or in other words, that tourism is not the centre of rural productive activities, as it can be a driving force for development, a complement or simply a one-off activity (p.7).

In the report "Rural tourism in Ecuador", prepared by Dr. Raquel Santos-Lacueva, in 2020. She determines that, in the field of rural tourism, consolidated experiences coexist, such as those of community tourism, and proposals recently implemented by the Ministry of Tourism. Thus, as demonstrated by the National Tourism Plan 2030, the Ecuadorian government continues to be committed to strengthening tourism development in rural areas, with community experiences, but also creating new products such as the revitalisation of localities through the implementation of the Magic Towns programme. Sanagustín (2018) determines that, "Rural tourism is characterised by the development in small territories with their own identity that have an extensive offer of diffuse, non-concentrated and small-scale accommodation and leisure activities".

The rural parish of Puerto Cayo in the canton of Jipijapa is a locality that has the appropriate geographical and socio-cultural characteristics for sustainable rural tourism, as it is possible to carry out activities related to this important tourist modality. One of its main tourist resources is the Cantagallo Protected Forest, where different activities can be carried out, such as: observation of the flora and fauna of the area and the preparation and tasting of the authentic "native" Montubio gastronomy (Gutiérrez, 2021).

The above shows the great possibilities for an alternative tourism business in the rural parish of Puerto Cayo in the canton of Jipijapa in the province of Manabí. However, in view of the new tourism trends in the world, which focus on environmental conservation practices and social support, and a greater awareness on the part of the population, it is essential that new rural tourism business projects have a product strategy based on a clear concept of sustainability.

The rural parish of Puerto Cayo in the canton of Jipijapa has a considerable tourist potential, which is not effectively exploited in a sustainable manner. However, the place has the necessary characteristics and qualities to be promoted through sustainable rural tourism, but there is a lack of proposals to make this possible. On the other hand, the lack of planning on the part of the competent authorities delays the development of tourism in the community. For this reason, it is important to mention the importance of having an action plan for the sustainable use of this modality. This research is especially aimed at rescuing the identity of the Montubio people of the rural parish of Puerto Cayo in the canton of Jipijapa, and subsequently promoting them as part of a sustainable rural tourism enterprise.

INDEX

I. LITERATURE REVIEW

This proposal requires the establishment of theoretical bases that arise from the identification of the variables of the research topic, and for this reason the terms that will underpin it are shown below.

1.1. Background

Tourism consists of several modalities that are increasingly adaptable to the environment in which it is implemented and in turn have specific characteristics that are appealing to different types of tourists, i.e. tourism is beginning to diversify, thus creating a number of possibilities when implementing modalities in specific destinations, therefore, it is necessary that the authorities who are responsible for the tourism sector, open their minds, investigate and explore the various destinations they have and in turn create projects to implement tourism modalities that generate economic, social and cultural benefits, It is therefore necessary that the authorities in charge of the tourism sector open their minds, investigate and explore the different destinations they have and in turn create projects to implement tourism modalities that generate economic, social and cultural benefits to the localities, taking into account that the synergy between managers and service providers creates a working environment in which both parties will benefit.

Through the journal Scielo, the researcher Mikery et al. (2014) states that the integration of tourism into rural productive activities represents a strategy to improve the living conditions of the inhabitants, especially because it is Research needs to start from a clear conceptualisation of the elements of rural tourism and the objective of rural tourism to guide the power analysis.

In a publication in the journal Mendive, Darias et al (2016). expresses that rural tourism needs community participation to ensure its sustainability over time and the achievement of its main objectives: customer satisfaction and local development with a positive impact for the community. Popular education, research, action and participation together make possible the direct participation of the community in identifying their needs, making decisions and designing possible solutions.

Rural environments where these activities take place are characterised by: low population density, landscapes and territories where agriculture prevails, and traditional lifestyles (UNWTO, 2019). The breadth of this definition makes it possible to include a wide range of tourism offerings within rural tourism. For example, one can speak of ecotourism, green tourism, agro-tourism, community-based tourism, cultural tourism, adventure tourism.

In August 2018, the Ministry of Tourism of Ecuador signed a "Cooperation Programme in Tourism Matters with the Ministry of Tourism of Mexico", with the purpose of transferring processes and methodologies to implement the Programme for the Development of Magical Towns in Ecuador. This initiative aims to fostering local development by promoting tourism development in localities with unique cultural and natural resources, and which meet certain preconditions for the development of tourism activity.

1.2. Theoretical basis Tourism.

Parra Oncins (2020) states that: "Tourism, despite all these considerations, under the paradigm of a sustainable, governed and ordered activity, can be a powerful instrument for the creation of businesses and the generation of quality employment".

Ledhesma (2016) considers that tourism is born from the displacement of the person, to a place other than his or her residence for the purpose of leisure, recreation or rest. On the other hand, tourism encompasses not only historical and environmental activities, but also artistic and educational activities in which the inhabitants and visitors at the destination are involved. Tourism today has given much to talk about, in the way in which it has been evolving and the ways of carrying out activities, in addition, tourism is not only focused on travellers who move from one place to another either for leisure, visits to relatives or business reasons. It also plays an important role in the planning, management and, therefore, development in territories where these services are provided and activities are practised.

Alternative Tourism

The Mexican Ministry of Tourism, (2004) affirms that Alternative Tourism is the reflection of this change of trend in the world, representing a new form of tourism, which allows mankind a re-encounter with nature, and a recognition of the value of interaction with rural culture, and at the same time, an opportunity for Mexico to participate in the segment with the greatest growth in the market in recent years.

Aguirre, Arroyo, & Navarro (2018) Therefore, alternative tourism responds to the needs and expectations of tourists seeking new experiences, reconnecting with nature and interacting with the host community's own culture. But this not only seeks environmental sustainability, but also a positive impact on the economic sphere that leads to sufficient income for all and that in turn is equitable, as well as the development and well-being of the local community in general.

Ibáñez & Rodríguez, 2012 It is a current of tourism that aims to carry out trips where tourists participate in recreational activities in contact with nature and the cultural expressions of rural, indigenous and urban communities, respecting the natural, cultural and historical heritage of the place they visit. This type of tourism is made up of activities that in its The name indicates its main characteristic: cultural tourism, rural tourism, agrotourism, ecotourism, adventure tourism, hunting tourism, among others. According to what has been mentioned, tourists are increasingly adapting to the different ways of doing tourism, in this case, they would get a phenomenal experience according to everything included, such as getting to know more about the natural environment, being entertained not only with flora and fauna activities but also taking advantage of and experiencing all the activities that are carried out within alternative tourism.

Local Development

Martinez (2010) asserts that, "Local development can be considered as the materialisation of solidarity behaviour between individuals willing to put their physical and financial resources to good use. These actions allow the population to satisfy their needs while exercising a certain degree of control over their future".

For Vicente & Morales (2005), local development is the set of results that are manifested in the improvement of the standard of living and quality of life of the inhabitants of a locality as a result of generating sustainable growth at various levels, which interlock, link, involve and complement each other in a strategic manner, capable of creating local improvement synergies that involve changing the systemic and structural conditions of the locality, deepening in the long term to the extent that a basic endogenous nucleus is formed and strengthened. According to Narvaez (2014) local development is not only based on an economic perspective, but also encompasses broader dimensions, involving the use of local capacities, which allow for a substantial improvement in the living conditions of the population. Framed in the concepts of several authors, local development is defined as the result of a synergic work between stakeholders, development that varies according to the sector it is focused on, but always keeping the same purpose, as tourism activity has positive and negative aspects, the way it is practised will depend on the social, cultural and natural impact on the different destinations.

1.3. Conceptual Framework Sustainable Rural Tourism

According to the Mexican Ministry of Tourism (2004), rural tourism is the most natural form of tourism, as it maintains the cultural identity and preserves the traditions and customs of rural communities. According to Fernández (2014), rural tourism can be understood as a tourist practice in rural areas that favours the economy and quality of life, through the offer of accommodation and leisure activities, with the mediating presence of the rural inhabitant, and which introduces the visitor to a living reality, with all its natural and cultural richness. UNWTO (2019) states that rural tourism is a type of tourism activity in which the visitor experience is related to a broad spectrum of products usually linked to nature activities, agriculture, rural lifestyles and cultures, angling and sightseeing. In the end, sustainable rural tourism remains that essential alternative that focuses on the rural environment, where the tourist can connect with the natural environment and engage in authentic site activities such as learning about agricultural processes and the cultural ways of life that communities may have within these sites.

Offer

Raffino (2020) states that, supply is defined as, the goods and services that different companies or individuals can sell in the market, in order to meet the needs of their users or consumers.

Tourist Offer

Socatelli (2013) defines tourism supply as the set of products and services associated with a specific geographical area, whose objective is to take advantage of the tourist attractions of that place, and whose Sellers can be offered on the market at a given price and within a given period of time, to be consumed by tourists. For Lemos (2003) "Tourism supply is the quantity of goods and services that a company (or set of companies) is able to produce and place on the market at a certain price, with a certain quality, in a certain place and for a certain period of time".

Demand

According to Raffino (2020), demand, in economics, refers to the quantity of goods or services that the population acquires to cover their needs or desires. These goods or services are very varied, from food, means of transport, education, leisure activities, medicines, among many other things, due to this it is considered that practically all human beings are demanders.

Tourism Demand

Ponosso & Lohman (2012) determine that, tourism demand is the total number of people participating in tourism activities, quantified as the number of tourist arrivals or departures, money spent during their stay and other statistical data. The main factors influencing tourism demand include the economic power of tourists, time availability and other motivating factors. Socatelli (2013) presents tourism demand as the set of consumers and potential consumers of tourism services seeking to satisfy their travel needs. These are tourists, travellers and visitors, regardless of the motivations that encourage them to travel and the place they visit or plan to visit.

Action Plan

According to Merino (2009), an action plan is a type of plan that focuses on the most important initiatives to achieve certain objectives and goals. In this way, an action plan is established as a guide that proposes a structure for carrying out a project. Reyna (2020) also states that an action plan is an administration or management tool, which, when implemented, will not only help to streamline processes, but also to outline clear goals and the best way to achieve them. Therefore, in the long term, this tool determines the tasks and resources used for the realisation of a plan that will lead a project to the proposed objective.

II. METHODOLOGY

The following methodological design was used to carry out the research work, in order to meet the objectives set out.

2.1. Theoretical Methods Descriptive-analytical method

The situational analysis of tourism in the locality was carried out using the SWOT matrix, based on an interview with the director of tourism of the decentralised autonomous government of the rural parish of Puerto Cayo.

Bibliographic method

By means of this method, different bibliographic sources were analysed, including the national tourism cadastre, elaborated by the Ministry of Tourism of Ecuador in 2020, which made it possible to determine the tourism offer of the rural parish of Puerto Cayo. In the same way, this method made it possible to determine the tourism demand of the rural parish of Puerto Cayo, since the research "Study of the tourism demand of the rural parish of Puerto Cayo", carried out during the first quarter of 2020, was used.

Deductive method.

This method was based on general premises about sustainable rural tourism in order to identify rural tourism activities that can be developed in the rural parish of Puerto Cayo.

Research Method.

The research method is a logical process through which knowledge is obtained. According to Ana Beatriz Ochoa G., it is a kind of compass in which knowledge is not automatically produced, but which prevents us from getting lost in the apparent chaos of phenomena, if only because it shows us how not to pose problems and how not to succumb to the spell of our favourite prejudices. The method, regardless of the object to which it is applied, aims to solve problems. The descriptive research method was used, according to Hernández (2017), who defines it as "a type of research that seeks to specify properties, characteristics and important features of any phenomenon to be analysed". The aim of descriptive research is to gain insight into prevailing situations, customs and attitudes through the accurate description of activities, objects, processes and people. Its goal is not limited to the collection of data, but to the prediction and identification of relationships between two or more variables.

2.2. Research Techniques

The main research techniques used in this work include the following:

Observation

This technique facilitated the situational analysis of the locality, since several visits were made that allowed for an accurate perception of the reality of tourism in the rural parish of Puerto Cayo.

Interview

This technique was used in order to obtain information related to the tourism situation in the parish, which was provided by the local tourism director. The present research is descriptive and was carried out in the rural parish of Puerto Cayo, Canton Jipijapa in the Province of Manabí, in which a market survey was used to measure the number of visitors who will come to learn about and enjoy sustainable rural tourism, whether they will be foreigners or nationals, in order to define the target public.

Survey

The survey, according to López-Roldán & Fachelli (2021), is used as a research method in which multiple specific techniques are involved in a coordinated way:

the design of the sample, the construction of the questionnaire, the measuring and the construction of indices and scales, interviewing, coding, organisation and monitoring of fieldwork (p. 9). The survey was used to obtain information on the opinion of 30 local people, including business people and tourism professionals, which allowed for a comparison of the results obtained in different ways, which was completed and made it possible to achieve greater precision in the information collected.

Sample

The sample, as indicated by Hernández & Carpio (2019), is defined as the subset of the universe or a representative part of the population, formed in turn by sample units that are the elements that are the objects of study, it is supported by sampling as a tool of scientific research whose main purpose is to determine the part of the population to be studied (p. 76). A sample of 30 people was used and 3 questions were asked to each individual, to obtain the information required for the objective of this research, the questionnaire was applied as an instrument to entrepreneurs and tourism professionals (in a personalised way and by telephone), in order to collect the information necessary to respond to the problem posed. (See in nterpretation of results, p. 31) The level of acceptance for a new approach to sustainable rural tourism in the rural parish of Puerto Cayo was measured with the questions asked, giving a positive result of 66.2 % for this proposal.

2.3. Sources of information.

Document that provides information for the study of a subject / Place where a flow of messages originates / In cataloguing, any printed item that can provide the information sought. It is any resource that responds to a user's demand for information, including information products and services, people or networks of people, computer programs, etc. (Course ha-2077 module III). According to María Dolores Ayuso, "Information sources and bibliographies produce ordered information either through references or documents, or through information from primary or secondary documents, and make it available and accessible to users".

Primary Sources

"They constitute the objective of bibliographic research in the literature review and offer first-hand data" (Hernández, 2017). According to Pedro Venegas: "They constitute the data that the researcher has available for analysis and that are the end product of his work through the application of any of the following techniques normally used, such as: observation, interview, experiment, application of censuses or surveys, among others. The information was obtained directly from the source, with the participation of people from the community: farm owners, tourism entrepreneurs, among others.

Secondary Sources

Defined by Pedro Venegas as all information, regardless of its nature, that the researcher has used, but that has not been generated by his work, in other words, information or data that precedes his work, should be considered as secondary source information, examples of this are: the documentary research process, literature or previous research that serves as a starting point, guide or support for his current research work. A minimum of processed information was obtained, such as: internet, brochures, bibliographic documents about sustainable rural tourism and thesis works and articles related to the topic, among others.

III. ANALYSIS AND TABULATION OF RESULTS

Objective 1: Situational analysis of tourism in the rural parish of Puerto Cayo.

Through an interview with the director of tourism in the rural parish of Puerto Cayo (see Annex 1) and constant visits to the site, extremely important information was obtained regarding the current tourism situation in the parish, which allowed for the creation of this SWOT matrix.

Table 1. SWOT analysis

SWOT ANALYSIS	
Strengths	Opportunities
The place has the necessary characteristics and qualities for the applicability of sustainable rural tourism proposals. High potential for sustainable rural tourism activities. Geographical location and favourable climate.Proximity to major economic cities and especially to international ports and airports (Manta, Guayaquil)	Creation of strategic agreements with tour operators outside the parish to attract new visitors. Short distance between tourist resources (it will favour the creation of routes, tourist circuits, etc.). Possible positioning as one of the most important rural destinations in the south of Manabí.
Weaknesses	Threats
•Sparse operation tourist at theparish. •Little exploitation of rural tourism potential. •Seasonality of demand. •No professionalism on the part of the managers of tourism activities in the area. Missing of innovation of tourism services	•Proximity to other potential destinations, such as: Puerto Lopez (Agua Blanca), Manta (Pacoche) •Low awareness of tourist attractions among travellers.

Source: Interview with the director of the Tourism Department of the GAD parish of Puerto Cayo and field work.

Prepared by: Marcos José Gutiérrez Bravo

Objective 2: Tourism supply and demand in the rural parish of Puerto Cayo.

3.1. Tourist resources of the rural parish of Puerto Cayo.

Table 2. Cantagallo Protected Forest.

Cantagallo Protected Forest	
Category:	Natural Attractions
Type:	Forest
Subtype:	Transition forest

Photograph:

Description of the attraction: The predominant vegetation is dry and humid forest with species such as Ceibo, Fernán Sánchez, Dormilón, Acacia, Balsa, Laurel, Guachapilí, Porotillo. They are permanently visited by howler monkeys (Alouatta palliata), a species of monkey that emits a characteristic howl that can be heard from 8 kilometres away.

Source: PDOT 2015-2019

Prepared by: Marcos José Gutiérrez Bravo

Table 3. Olina Forest.

Olina Forest	
Category:	Natural Attractions
Type:	Forest
Subtype:	Humid Tropical

Photograph:

Description of the attraction: It is a humid forest of garúa, with very rich fauna and flora, there is a natural water collector called "El Chorrillo" that collects along the mountain the water of the superior part, that is product of the fog that covers the superior vegetation and, that filters to the secondary collectors. This attractive forest is composed of trees, shrubs and creeping vegetation, products of planting mounds and grasslands, with trees that reach up to 25m. Variety of bromeliads and ferns.

Source: PDOT 2015-2019

Prepared by: Marcos José Gutiérrez Bravo

Table 4. Pre-Hispanic stone sculpture from El Barro-Cantagallo.

Pre-Hispanic stone sculpture from EL Barro-Cantagallo	
Category:	Cultural Attraction
Type:	Architecture
Subtype:	Archaeological area
Photograph: 	
Description of the attraction: Despite abandonment, weathering and natural modifications, the general outline of the plan of the structure has not been modified. It consists of superimposed stones, of variable size, weight and shape, although of varied materials (sandstone, calcareous conglomerates), arranged longitudinally forming "walls" of the type of "gabions", similar to those of the from the Agua Blanca site, Manteño style.	

Source: PDOT 2015-2019

Prepared by: Marcos José Gutiérrez Bravo

Table 5. Cabuyeros of Manantial.

Cabuyeros de Manantial	
Category:	Cultural Attraction
Type:	Popular and cultural heritage
Subtype:	Crafts and arts
Photograph:	

Description of the attraction: At the Manantiales site, the native plant "cabuya" is still harvested and after a process of debarking, washing and drying, lots are formed which are then stacked and sold. However, nowadays the transformation into ropes and fabrics has disappeared, they only harvest, debark and sell the bundle to artisans in Montecristi for two dollars. dollars per quintal. These handicrafts are used as cordage and weavings.

Source: PDOT 2015-2019

Prepared by: Marcos José Gutiérrez Bravo

Table 6. Puerto Cayo Beach

Puerto Cayo Beach	
Category:	Natural Attractions
Type:	Coasts or coastlines
Subtype:	Beach
Photograph: 	
Description of the attraction: The beach of Puerto Cayo is currently one of the most beautiful and exclusive beaches in Ecuador, with a total extension of 12 kilometres of white sand and dunes bordered by Cape San Jose, its geographical location makes it the owner of a pleasant climate all year round, not to mention being one of the most privileged places for birdwatching, and one of the most beautiful and exclusive beaches in Ecuador. whales between June and September.	

Source: PDOT 2015-2019

Prepared by: Marcos José Gutiérrez Bravo

Table 7. Islet Pedernales.

Pedernales Islet	
Category:	Natural Attractions
Type:	Island Lands
Subtype:	Islet
Photograph: 	
Description of the attraction: This rock formation of volcanic origin is surrounded by steep cliffs and a beautiful turquoise sea due to the enormous amount of white coral on its seabed, on the other hand, its surface covered by tropical dry forest serves as a habitat for several species of birds, such as: pelicans, a small colony of blue-footed boobies and frigate birds.	

Source: PDOT 2015-2019

Prepared by: Marcos José Gutiérrez Bravo

Table 8. Beach Boca de Cayo

Boca de Cayo Beach	
Category:	Natural Attractions
Type:	Coasts or coastlines
Subtype:	Beach
Photograph:	
Description of the attraction: The beach of Puerto de La Boca de Cayo is located in the northern part of the parish. It has mangroves in the immediate vicinity and a pleasant current that makes the sea very calm for bathers.	

Source: PDOT 2015-2019

Prepared by: Marcos José Gutiérrez Bravo

3.2. Tourist plant in the rural parish of Puerto Cayo.

Table 9. Dreams from sea

Name of establishment	Classification	Category	Rooms	Camas	Plazas
Dreams of the sea	Hosteria	Third	31	93	93
Contact: Telephone: 080983022 					

Source: National Tourism Cadastre 2020

Prepared by: Marcos José Gutiérrez Bravo

Table 10. The Cabaña

Name of establishment	Classification	Categorieía	Roomis	Camas	Plazas
The hut	Hosteria	Third	6	6	12
Contact: Telephone: 052616030					

Source: National Tourism Cadastre 2020

Prepared by: Marcos José Gutiérrez Bravo

Table 11. Sanctuary Puerto Cayo Lodge

Name of establishment	Classification	Category	Rooms	Camas	Plazas
Sanctuary Puerto Cayo Lodge	Hostel	3stars	13	37	37
Contact: Mobile: 0999483709 Email: mdager@hotmail.com Web: www.sanctuaryecuador.com					

Source: National Tourism Cadastre 2020

Prepared by: Marcos José Gutiérrez Bravo

Table 12. Cabalonga Eco- Aventure

Name from establishment	Classification	Category	Roomis	Camas	Plazas
Cabalonga Eco Adventure	Campamento Tourist	Category a Unica	7	11	14
Contact: Telephone: 052387149 Mobile: 0995375888 Mail: cabalongaecoadventure@gmail.com Web: www.cabalonga.com					

Source: National Tourism Cadastre 2020

Prepared by: Marcos José Gutiérrez Bravo

Table 13. The Tanusas

Name from establishment	Classification	Categorya	Roomis	Beds	Places
The Tanusas	Hosteria	First	13	13	39
Contact: Phone: 023962900 Email: info@lastanusas.com Web: www.lastanusas.com 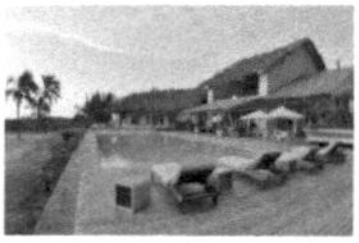					

Source: National Tourism Cadastre 2020

Prepared by: Marcos José Gutiérrez Bravo

Table 14. Light from Luna

Name from establishment	Classification	Categorya	Roomis	Beds	Places
Moonlight	Hosteria	Second	33	97	120
Contact: Telephone: 052616031 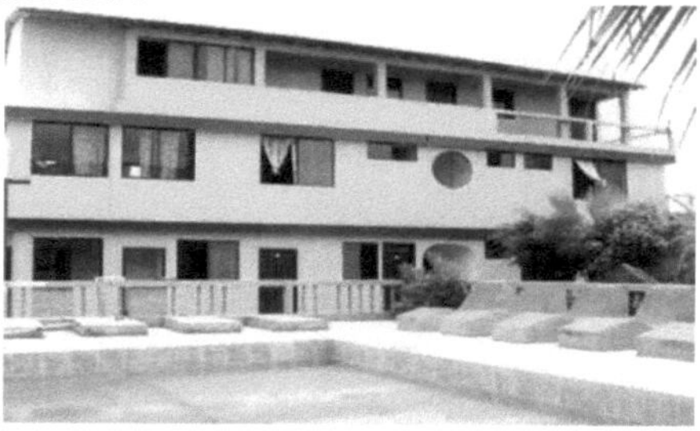					

Source: National Tourism Cadastre 2020

Prepared by: Marcos José Gutiérrez Bravo

Table 15. Catering establishments in the rural parish of Puerto Cayo.

Establishment	Ranking	Tables	Places	Address
Pepita's Creole flavour	Cabaña Restaurant	7	28	Main street next to the petrol station
Dianita	Cabaña Restaurant	10	40	Av. Guayas and Malecón
Coco Beach	Cabaña Restaurant	5	20	Av. Guayas and Malecón
Nicolle	Cottage Restaurant	6	24	Av. Guayas and Malecón
Nicolle 2	Cabaña Restaurant	4	16	Av. Guayas and Malecón
Golden Seal	Cabaña Restaurant	12	48	Av. Guayas and Malecón
El Gringo	Restaurant	20	80	Seawall next to the naval checkpoint
Barandhua	Restaurant	10	40	Av. Guayas and Malecón
Don Carlos	Dining Bar	10	40	Malecón next to Zavala's hostel
Caprice	Restaurant	10	40	Av. Guayas and Malecón
Margarita	Restaurant	7	28	Av. Guayas and Malecón
Katherine	Restaurant	5	20	Guayas and Simón Bolívar Av.

Source: National Tourism Cadastre 2020

Prepared by: Marcos José Gutiérrez Bravo

Analysis of the tourist offer of the parish: The tourist resources of the rural parish of Puerto Cayo are extremely important for the local tourist development, where the Pedernales islet, the Puerto Cayo beach and the Cantagallo protective forest stand out and provide that distinctive touch to the place, as well as having other sites of tourist interest, such as the recently built boardwalk, where In the same way, it is important to mention that several places in the locality, such as the communes of Cantagallo, La Boca, Olina and Manantiales, have the necessary qualities to carry out tourist activities linked to sustainable rural tourism, a type of tourism that can boost the parish by proposing this type of activity. In short, from rainforests to tropical paradises, the tourist potential of the parish is unique and should be exploited sustainably in tourism.

3.3. Tourism demand in the rural parish of Puerto Cayo.

This outline of the characteristics of tourism demand in the rural parish of Puerto Cayo was obtained through the analysis of the document "Study of tourism demand in the cantons of the province of Manabí: case study of the canton of Jipijapa - rural parish of Puerto Cayo", prepared by applying surveys to 30 local people, including business people and tourism professionals, who were in the parish during the first quarter of the year 2024.

Table 16. Characteristics of the profile of tourists visiting Puerto Cayo.

Profile of the tourist visiting Puerto Cayo		
Source		
Manabí	197	44%
Guayas	165	37%
St. Helena	5	1%
Pichincha	12	3%
Cotopaxi	6	1%
Other domestic destinations	45	10%
Argentina	4	1%
Other international destinations	16	4%
Age Range		
Age	18-50	
Gender		
Male	50%	
Female	45%	
LGBT	5%	
Marital Status		
Single	57,33%	
Married	27,11%	
Another	15,56%	
Occupation		
Trader	12%	
Housewife	12%	

Nurse	6%
Student	26%
Another	44%
Reasons for travel	
Rest	56%
Recreation	26%
Visiting family/friends	7%
Other personal reasons	11%
People you travel with	
Only	9%
1-2 Persons	13%
3-4 Persons	18%
5+ Persons	60%
Length of stay	
Less than 1 day	46%
1-2 days	33%
3-4 days	12%
5-6 days	9%
Factors influencing the visit	
Recommendations	31%
Proximity to the place of origin	20%
Time availability	12%
Prices	11%
Interest in getting to know the place	2%
Work	13%
Visiting family and friends	2%
Prior knowledge	0%
Diversity of activities	7%
Event advertising	2%
Types of accommodation establishment used during the stay on site	
Hotel	9%
Hostel	14%
Pension	1%
Family home	20%
Own house or flat	1%
House or flat for rent	4%
Types of catering establishment used during the stay on site	
Cabins	79%
Barbecues	2%
Restaurants in accommodation	3%

Speciality restaurants		17%
Approximate daily expenditure per service		
Restoration	10 a 20	34%
	30 a 40	43%
	50 a 85	23%
Accommodation	10 a 20	12%
	30 a 40	64%
	50 a 85	24%
Souvenirs	10 a 20	56%
	30 a 40	14%
	50 a 85	30%
Recreation	10 a 20	74%
	30 a 40	18%
	50 a 85	8%

Source: Study of tourism demand in the cantons of the province of Manabí:

case study Jipijapa canton - rural parish Puerto Cayo

Prepared by: Marcos José Gutiérrez Bravo

Analysis of the tourist profile: In relation to the place of origin of the tourists arriving in the rural parish of Puerto Cayo, 39% are tourists from Manabí, followed by 28% from the province of Guayas, 10% from the province of Pichincha, 3% from the provinces of Santa Elena and Cotopaxi, while 20% are international tourists. Meanwhile, the age range is between 18 and 60 years old. Similarly, the gender of the same is mostly male with 50%, compared to 45% female, and to a lesser extent the LGBT segment with 5%, which shows that the site is not very popular with the latter mentioned. Continuing with marital status, 57.33% are single, 27.11% are married and 15.56% are divorced or widowed, among others. In reference to occupation, we have shopkeepers with 12%, housewives with 12%, nurses with 6% and other professions with 70%. The travel motivations presented by the travellers are mainly rest, with a percentage of 56%, followed by recreation with 26%, visits to family and friends with 9% and other personal reasons with 9%. It is also shown that tourists travel in the company of more than five people with a total percentage of 60% and 40%

travel alone. In turn, the number of days of stay in the place is established as follows: 46% of people staying less than one day, followed by 33% who stay between 1-2 days, 12% between 3-4 days and 9% 5-6 days. Following with the factors influencing the visit, we have recommendations (word of mouth) with 51% as the dominant percentage, followed by proximity to the place of origin with 34% and to a lesser extent to advertising in events with 15%. It was identified that the majority (53%) of tourists arriving in the area spend the night at their relatives' homes, while the most frequently used accommodation establishment is the hostels, with 47%. At the same time, the most popular catering establishments are cabins (79%) and other establishments (21%) and, finally, the approximate daily expenditure per service is determined as follows: In catering services, the dominant average expenditure is 30-40 $ with 43%, in accommodation services in the same way the expenditure between 30-40 $ stands out with 64%, in souvenirs 56% expenditure between 10-20 $ and in recreation services, the highest expenditure is 10-20 $ with a percentage of 74%.

Objective 3: Rural tourism activities that can be developed in the parish of Puerto Cayo. In the following table, activities linked to rural tourism are proposed.

The project aims to develop sustainable rural tourism activities that can be carried out in several of the communities of the rural parish of Puerto Cayo, taking into account their own characteristics that facilitate the implementation of these activities, thus enabling tourism in the parish of Puerto Cayo to be boosted through sustainable rural tourism as an applicable modality for tourism development.

Table 17. Activities linked to sustainable rural tourism

Activity	Site where you can to carry out the activity	Description of the activity
Ethnotourism	Communities: La Boca de Cayo, Olina, Cantagallo and Manantial.	Tourist activity carried out with the aim of learning about the customs and traditions of a given region. culture.
Preparation and tasting of gastronomy traditional	Communities: Olina and Cantagallo	The aim of this tourist activity is for tourists to learn about the ancestral preparation and to taste the variety local gastronomy.
Agritourism	Communities: La Boca de Cayo, Olina, Cantagallo and Manantial.	Through this activity, tourists have the opportunity to experience the livestock farming practices, agricultural crops and the way of life of the inhabitants of these rural communities.
Mystical experiences	Communities: Olina and Cantagallo	This activity offers the opportunity to get to know the richness of the beliefs and legends of the communities.
Eco Archaeology	Communities: El Barro-Cantagallo (Pre-Hispanic stone sculpture)	It is based on the interest in learning about the relationship between man and his environment from the vestiges materials.
Preparation and use of traditional medicine	Communities: La Boca de Cayo, Olina, Cantagallo and Manantial.	This experience provides an opportunity to immerse oneself in the world of traditional medicine that rural communities have maintained for generations.
Workshops crafts	Community of Manantial (Cabuyeros)	Thanks to this activity, tourists Learn theprocedure at
		different types of handicrafts and the respective processing needed for the materials with which they are made. elaborate.
Rural photography	Communities: La Boca de Cayo, Olina, Cantagallo and Manantial.	Tourists have the opportunity to photograph the beautiful scenery of these rural locations and their natural bounties. cultural.
Flora and fauna sightings	Communities: Olina and Cantagallo (Monkey observation howlers)	This activity allows you to witness the natural environment and the different plants and animals that live in the area. its immediate vicinity.
Equestrian rides	Puerto Cayo beach and La Boca beach	Short rides on the beach are offered for short periods under the respective prevention measures.of time.

Prepared by: Marcos José Gutiérrez Bravo

3.4. Results and discussion Development of instruments.

In the survey and sample technique: a questionnaire was used as a research instrument, which allowed questions to be asked of local people, including business people and tourism professionals. In the direct observation technique, the camera was used as the main instrument.

Application of information tools and processes.

In the application of the survey technique, a questionnaire was used as an instrument, for which a small sample of the population was taken to gather information that allowed us to draw up a brief inventory of the natural and cultural resources.A sample of 30 local people was taken, including entrepreneurs and tourism professionals from the rural parish of Puerto Cayo, which allowed information to be obtained about the level of acceptance of the proposed model for a new approach to sustainable rural tourism. In order to make the research more precise, the customs and culture of the rurality of the site were introduced, where the observation technique was applied, in order to investigate, dialogue and be in the field of action.

The profile of tourists entering the rural parish of Puerto Cayo was also studied by means of direct observation and research at ITUR Jipijapa.

Interpretation of results.

For the fulfilment of the specific objectives, the results contained in the statistical tables obtained in the surveys were considered and are presented as follows:

Question 1

Do you think that a proposal for rural tourism should be made under the precepts of sustainable development that would allow visitors to have another alternative?

Table 18. Proposal for sustainable rural tourism

Response & total	Persons	Percentage
Yes	29	96.6
No	1	3.4
Total	30	100

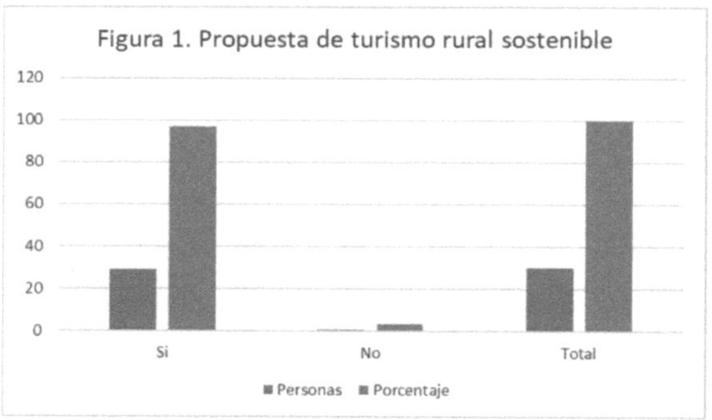

Figure 1. Proposal for sustainable rural tourism

Question 2

Do you think that sustainable rural tourism should be promoted and developed, in which the objectives of good customer service and total quality are achieved, without generating alterations and harmful impacts on the environment?

Table 19. Promotion and development of sustainable rural tourism

Response & total	Persons	Percentage
Yes	27	90
No	3	10
Total	30	100

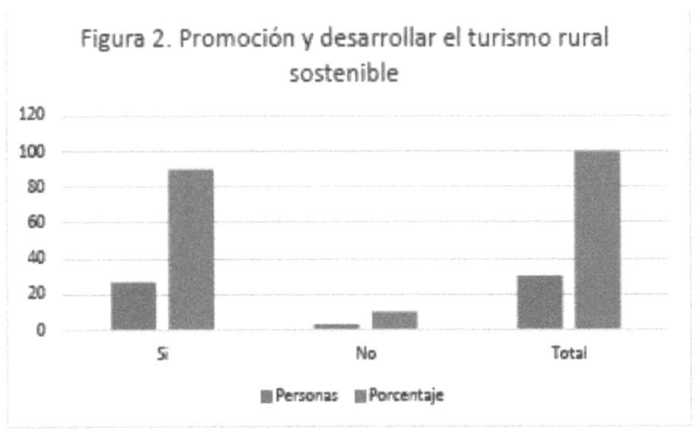

Figure 2. Promoting and developing sustainable rural tourism Question 3

Do entrepreneurs have the necessary tools (regulations, policies, programmes, etc.) to enable them to carry out a tourism activity that is well oriented and channelled towards customer satisfaction, profit generation and the protection of the natural environment?

Table 20. Entrepreneurs have the necessary tools

Response & total	Persons	Percentage
Yes	27	12
No	3	88
Total	30	100

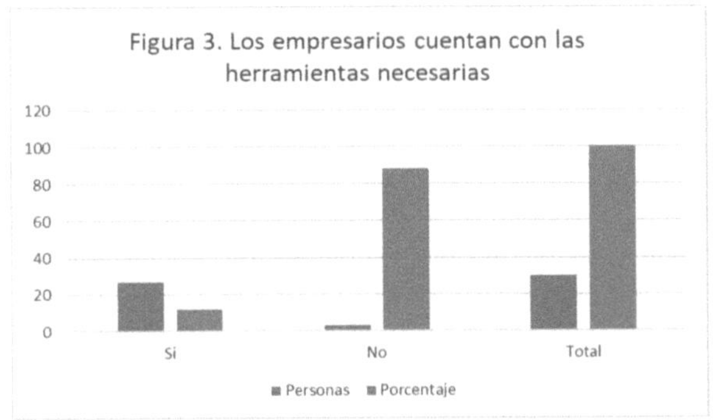

Figure 3. Entrepreneurs have the tools they need

IV. DISCUSSION

Through the research to achieve the specific objectives, for which the results were obtained from the statistical tables which in turn come from the survey made to entrepreneurs and professionals in the tourism field, it brings to light that: the level of acceptance of the proposal is 66.2 %, which suggests that the proposal will be successful for entrepreneurs who wish to to venture into this modality. On the other hand, according to the statistics, 33.8% say that they do not have the necessary tools such as regulations, policies, programmes, etc. This is worrying and there is a need to work on this issue in order to achieve the proposed goal. This is worrying and there is a need to work on this issue in order to achieve the proposed goal. Tourism as a destination in the rural parish of Puerto Cayo in the canton of Jipijapa, is at the beginning stage with projections to grow in the coming years, due to its great natural, cultural, religious, handicraft and gastronomic potential. Statistics indicate the number of tourists who visit: approximately 15,000 per year, of which 80% are national and 20% are international. In 2009, according to data from the Ministry of Tourism (MINTUR), 968,499 tourists arrived in Ecuador, representing 100%, of which 1.573% visited. The parish offers the tourist a contrast of ancient and modern, hospitality of its people and different microclimates with beautiful landscapes.

V.CONCLUSIONS

The rural parish of Puerto Cayo in the canton of Jipijapa currently has a great potential for tourism, giving the population an opportunity to generate income through this activity, with non-mass tourism products that are environmentally friendly.Entrepreneurs need to have the necessary tools such as policies, programmes and regulations. This will allow them to develop and guide them on the limits of their tourism operation. There is a need for awareness-raising work on sustainable policies, which will require a great effort, for which the tourism entrepreneurs of the rural parish of Puerto Cayo must make joint efforts among themselves and establish alliances with those who have some kind of influence on the Montuvian people.Although the population has the potential in terms of tourism, natural, gastronomic and human resources, it is necessary to provide adequate training to all those involved in tourism management. Being sustainable will not necessarily make them profitable. They will need to develop appropriate marketing strategies on an ongoing basis in order to reach and attract the right audience in the shortest possible time. In the rural parish of Puerto Cayo in the canton of Jipijapa, as a tourist destination, it is in the initial or introductory stage, which is why it is necessary to insist with the following entities public bodies (e.g. Chamber of Tourism, Ministry of Tourism) to ask for support in promotion at national and international level. This investigation proceeded to carry out a meticulous study on the origins of the Montubian culture, applying research methods in accordance with the objectives. The result of this investigation was that the majority of people who consider themselves Montubians agree with the proposal. It was determined that the rural parish of Puerto Cayo has a high potential for sustainable rural tourism activities, as its geographic, cultural, climatic, economic and social qualities allow for the development of a large number of activities linked to this modality; however, the lack of planning on the part of the managers of tourism activities in the locality has impeded the development of these activities. On the other hand, the tourist offer of the locality was identified, where the natural tourist resources predominate, among the most outstanding and well-known are the beach of Puerto Cayo, the Pedernales islet and the Cantagallo protective forest, as well as the great variety of accommodation and catering establishments. In addition to distinguishing the profile of the tourists who visit the parish, where the majority are established as national tourists, coming from the province of Guayas, neighbouring towns in the province of Manabí and international tourists. Finally, it has been established that the natural and cultural assets of several localities in the parish allow for a number of activities linked to sustainable rural tourism, but that, due to the lack of promotion, planning and management in the implementation of these activities, the use and development of sustainable rural tourism in the sites addressed in the research has been delayed.

VI. PROPOSAL

4.1. Title of the Proposal

Propose an action plan for the development of sustainable rural tourism in the rural parish of Puerto Cayo.

4.2. Summary of the Proposal

The rural parish of Puerto Cayo has sites of tourist interest for the implementation of activities that allow for the promotion of the locality through sustainable rural tourism. For this reason, the present action plan seeks to promote the implementation of this modality. In the same way, tourism products located in the different rural communities throughout the parish are proposed.

9.1. Mission

Promote the practice of sustainable rural tourism in the rural parish Puerto Cayo.

9.2. Vision

In the year 2027, sustainable rural tourism will be one of the main tourism modalities that will drive local development in the rural parish of Puerto Cayo.

9.3. General Objective

Structure the action plan with strategies aimed at promoting tourism development in the parish of Puerto Cayo through sustainable rural tourism.

9.4. Specific Objectives

- Establish the phases for the elaboration of the action plan for the sustainable use of rural tourism in the rural parish of Puerto Cayo.
- Design an action plan for the sustainable use of sustainable rural tourism in the rural parish of Puerto Cayo.
- Creating a sustainable rural tourism product for the rural parish

Puerto Cayo.

9.5. Scope of the Proposal

Through the design of this action plan, the aim is to strengthen the practice of sustainable rural tourism in the rural parish of Puerto Cayo, as well as the initiative that this study will be used by the managers of tourism activities in the territory for future planning.

9.6. Methodology of the work

Specific Objective 1 Establish a plan of action for the sustainable use of sustainable rural tourism in the rural parish of Puerto Cayo.

Figure 4. Outline of the elaboration of the action plan

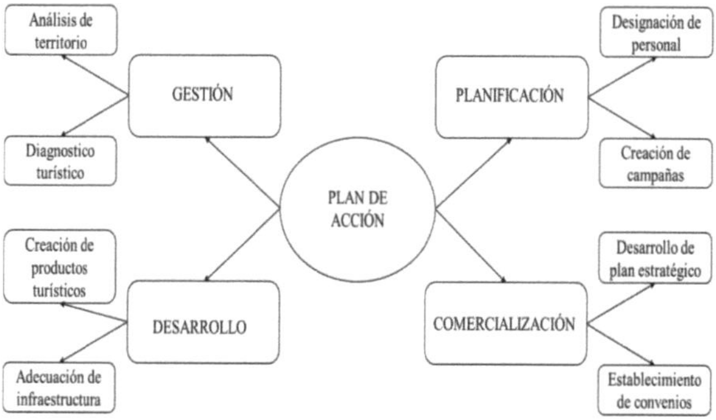

Prepared by: Marcos José Gutiérrez Bravo.

By means of this scheme, it was possible to determine the phases that will structure the action plan. The scheme is made up of the management, planning, development and marketing phases.

Phase 1: Management

Through the management phase, an analysis of the territory is carried out to determine the sites of tourist interest and the resources available in the locality, and a tourism diagnosis is carried out to determine the characteristics necessary for the implementation of the tourism modality.

Phase 2: Planning

The second phase is planning, in which personnel are appointed to manage and deliver the tourism activity, followed by the creation of training campaigns for the different communities that will be involved in the tourism activity.

Phase 3: Development

The third phase is the development, where tourism products will be designed, maintaining the concept of innovation and quality of service, as well as the adaptation of the infrastructure, through constant evaluations in the different facilities of the localities involved.

Phase 4: Commercialisation

In this phase, strategies will be implemented that will help to market the products previously developed, thus promoting the destination and the different communities that provide tourism services, as well as agreements with institutions involved in the tourism sector.

Specific Objective 2 Design an action plan for the sustainable use of sustainable rural tourism in the parish of Puerto Cayo.

Table 21. Action plan for the development of sustainable rural tourism in the rural parish of Puerto Cayo.

DIMENSION	AXES	ACTION	EXECUTOR	PERIODICITY
MANAGEMENT	RESEARCH	Carry out a diagnosis of the canton's tourism potential.	GAD parish of Puerto Cayo.	1 time within 6 months.
	RESEARCH	Create a record of the most visited places and other information of interest. for the tourist.	GAD parish of Puerto Cayo.	Monthly
	STRATEGIC MANAGEMENT	Implement a digital platform where all the tourist information of the place is compiled so that it is completely available and accessible for the inhabitants, tourists, the private sector and the public. general public.	GAD parish of Puerto Cayo.	1 time.
PLANNING	STRATEGIC MANAGEMENT	Appoint action and strategic management staff specialised in marketing, events, economic development to reinforce the competent procedures of the Tourism Directorate of the parish council of GAD of Puerto Cayo.	GAD parish of Puerto Cayo.	1 time.
	RESEARCH	Create environmental education campaigns to raise awareness in society and mitigate the negative impacts that tourism generates on the environment. natural.	MINTUR/ GAD parish of Puerto Cayo.	Every 3 months.

	AWARENESS RAISING IN THE TOURISM SECTOR	Execute tourism training programmes for the inhabitants of the rural communities of Puerto Cayo.	Ministry of Tourism/G A D parish of Puerto Cayo/UNESUM	Every 3 months.
	SUSTAINABLE TOURISM DEVELOPMENT	Implementation of an awareness-raising programme for tourism service providers on the creation of tourism-related products. rural tourism.	Ministry of Tourism/G A D parish of Puerto Cayo.	For 2 years.
	TOURISM PRODUCT AND EXPERIENCE	To conceive working meetings with local groups and relevant stakeholder bodies to provide advice for the design of rural tourism products where the tourism potential is fully integrated with the that the parish has.	GAD Puerto Cayo parish, Host community, private sector, consultants	For 6 months.
DEVELOPMENT	SERVICE AND QUALITY STANDARDS	Develop training campaigns for tourism workers in order to increase their skills and improve the level of customer service. client.	Ministry of Tourism.	Every 2 months.
	SUSTAINABLE DEVELOPMENT.	Working with local rural communities and agencies to promote traditional heritage practice and customs.	GAD parish of Puerto Cayo, Ministry of Tourism	For 2 years.
		Organisation and beautification of the public spaces.	GAD parish of Puerto Cayo.	1 year.

	INFRASTRU CTURE	Improving existing rural infrastructure and upgrading it to provide services. tourism.	GAD parish of Puerto Cayo.	1 year.
		Signposting and identification of rural areas and sites intended for rural use. tourism.	GAD parish of Puerto Cayo.	1 year.
	SERVICE AND QUALITY STANDARD S.	To carry out constant evaluations and reviews of tourism employees in order to maintain the optimal level of the quality of service.	GAD parish of Puerto Cayo and the Ministry of Tourism.	Monthly.
	SERVICE AND QUALITY STANDARD S.	Survey tourists and visitors to find out their level of satisfaction and the quality of the service provided during their stay in the parish.	GAD parish of Puerto Cayo.	Monthly.
	SUSTAINAB LE DEVELOPM ENT.	Train the community at large, local businesses, tourism service providers on the creation of tourism business models. sustainable.	Ministry of Tourism, GAD parish of Puerto Cayo.	Every 3 months.
	MARKETIN G STRATEGIC .	Develop a strategic tourism marketing plan.	GAD parish of Puerto Cayo.	3 months.

MARKETIN G.	STRATEGIC MARKETIN G.	Establish agreements with neighbouring localities, such as Puerto López, Manta, Portoviejo and Jipijapa to jointly develop a rural tourist route and efficiently take advantage of the tourism potential of the region. count.	GAD'S of the cantons of Puerto López, Jipijapa, Manta and Portoviejo,	Only once a year.
	INFORMATI ON SERVICES FOR VISITORS	Publicise the destination and its respective rural properties.	GAD parish of Puerto Cayo.	For 2 years.
	BRAND IMAGE.	To create a brand and identity for the parish that is attached to rural tourism, so that it can be distinguished and recognised. at the national level.	GAD parish of Puerto Cayo.	One time only.
	EVENTS.	Through the GAD's tourism department, organise expo-fairs and festivals to showcase the identity and potential of rural tourism. of destiny.	GAD parish of Puerto Cayo.	Semi-annually.

Prepared by: Marcos José Gutiérrez Bravo.

Specific Objective 3 Create a rural tourism product for the parish of Puerto Cayo.

This tourism product is shown as an option for the sustainable use of the rural communities of the parish, and it should be emphasised that it is based on the activities proposed in the specific objective number three of this research.

Figure 5. Tourism product

Prepared by: Marcos José Gutiérrez Bravo.

Figure 6. Tourist product Forests Route

Prepared by: Marcos José Gutiérrez Bravo.

Figure 7. Product description (Route of the Forests)

Prepared by: Marcos José Gutiérrez Bravo.

Figure 8. Tourist product Watermelons-Cabuya Route

Prepared by: Marcos José Gutiérrez Bravo.

Figure 9. Product description (Watermelon-Cabuya route)

Prepared by: Marcos José Gutiérrez Bravo.

VII. BIBLIOGRAPHY

Acerenza, M. (2003). Destination marketing management in today's competitive environment. Inputs and Transfers, 43-56.

Alvarez Alvarado, R. (2022). El turismo rural y el desarrollo local sostenible desde la percepción de los pobladores de la parroquia Ingapirca. Magazine Publishing,9(33), 67-86. https://doi.org/10.51528/rp.vol9.id2278 Available en:

https://revistapublicando.org/revista/index.php/crv/article/view/2278/25 03

World Bank, World Economic Outlook, June 2020, Washington D. C: World Bank.

Blain, C., Levy, S.E. and Brent Ritchie, J.R. (2005), "Destination Branding: Insights and Practices from Destination Management Organizations", Journal of Travel Research, vol.43, 328-338.

Boullón, R. C. (1985). Planificación del espacio turístico. Mexico D.F.: Trillas.

Butler, R. (1980). The concept of a tourist area cycle of evolution: Implications for management of resources. Toronto: Canadian Geographer.

Cooper, C. (2001). Turismo: principios e prácticas. Sao Paulo: Bookman.

Crosby, A. (2009). Re-inventing rural tourism: Management and development. Re-inventing rural tourism, 1-227.

Darias Fuertes, M., Ramírez Pérez, J., & Pérez Hernández, M. (2016). The development of rural tourism from the conception of Education.

Popular. Mendive. Journal of Education, 14(4), 308-313. Retrieved from https://mendive.upr.edu.cu/index.php/MendiveUPR/article/view/868

De La Rosa, B. (2003). La imagen turística de las regiones insulares: Las islas como paraísos. Cuadernos de turismo, 127-137.

Eby, D., L. Molnar, And L. Cai (1999), "Content Preferences for In-Vehicle Tourist Information System: An Emerging Information Source", Journal of Hospitality and Leisure Marketing 6(3), 41-58.

El Comercio (May 22, 2020) Ecuador's tourism sector will lose up to USD 400 million per month by the pandemic. Retrieved from: https://www.elcomercio.com/tendencias/perdidas-sector-turistico-ecuadorcoronavirus.html

Erdem, T., Swait J. (1998), "Brand Equity as a Signaling Phenomenon", Journal of Consumer Psychology, Vol. 7, No. 2, 131-158, 243-254.

Fernández, J. (2015). Economic boost through tourism in Sub-Saharan Africa. Asia and Africa Studies, 78-115.

Frutos, Lopez-Roig, Serra Cobo and Devaux (12 May 2020) COVID-19: The Conjunction of Events Leading to the Coronavirus Pandemic and Lessons to Learn for Future Threats. Frontiers in Medicine, 223. https://doi.org/10.3389/fmed.2020.00223

Gutiérrez, M. (2021) El Turismo Rural como Impulso Turístico de la Parroquia Puerto Cayo, Cantón Jipijapa, Provincia de Manabí. UNESUM. Faculty of Economic Sciences.

Hernández, S. (2017). Metodología de la investigación. Sixth Edition, Mexico: Editorial McGraw-Hill. Available at: https://www.uca.ac.cr/wp-content/uploads/2017/10/Investigacion.pdf

Hernández Ávila CE, Carpio N. Introduction to types of sampling. ALERTA Journal. 2019; 2(1): 75-79. DOI: https://doi.org/10.5377/alerta.v2i1.7535. Available at: https://alerta.salud.gob.sv/wp-content/uploads/2019/04/Revista-ALERTA-Year-2019-Vol.-2-N-1-vf-75-79.pdf.

Ivars, Joseph A., 2003: Tourism Planning, Spain: Síntesis.

Kotler, P. (1995). Marketing: Latin American Edition. Mexico: Pearson Educational.

López - Roldán, P., & Sandra, F. (2021). The Survey. Electronic edition. Full text at https://mdx.cat/handle/10503/105303

Mapelli, Giovanna 2008 Las marcas de metadiscurso interpersonal de la sección turismo de los sitios web de los ayuntamientos. In Calvi, MariaVittoria, Mapelli, Giovanna and Santos López, Javier (Eds.), Lingue, culture, economia: comunicazione e pratiche discorsive, Milano, FrancoAngeli, 173-190.

Mikery Gutiérrez, Mildred Joselyn, & Pérez-Vázquez, Arturo (2014). Methods for the analysis of the tourism potential of rural territory. Mexican Journal of agricultural sciences, 5(spe9),17291740. https://doi.org/10.29312/remexca. v0i9.1060

Monferrer, D. (2013). Fundamentals of marketing. Spain: UNE. Ojeda, M. A. (2013). User involvement in advertising production. A reflection on the spontaneous advertising generated by the users/consumers. Magazine ICONO, 14(1), 303-317. Doi: 10.7195/ri14.v11i1.204

UNWTO,"Figures from tourists International Tourist Figures" See Available at see at https://www.unwto.org/es/news/covid-19-las-cifras-de- international-tourists-could-fall-60-80-in-2020.

Ruiz, R. (2017). Participatory reactivation of public space. Culturas, Revista de gestión cultural, 93-116.

Santesmases, M. (2012). Marketing: Concepts y Estrategias.Madrid: Pirámide.

Tajada, S. D. (1974). The fundamentals of marketing and some types of commercial research. Madrid: ESIC.

Thomé, H. (2008). Turismo rural y campesinado, una aproximación social desde la ecología, la cultura y la economía, Mexico. Convergencia, Revista de

Ciencias Sociales.

Valverde Sánchez, R. Y. (2017). Plan de Promoción Turística Para El Incremento De La Afluencia De Turistas En El Refugio De Vida Silvestre Laquipampa - Incahuasi. January - September 2016.

William Perreault, J. M. (1996). Marketing. Mexico: D.F: Irwin.

World Tourism Organization (20 January 2020) International tourism growth continues to outpace the global economy. Retrieved from: https://www.unwto.org/international-tourismgrowth-continues-to- outpace-the-economy

World Tourism Organization (1 April 2020) Call for Action for Tourism COVID-19 Mitigation and Recovery. Retrieved from: https://webunwto.s3.eu-west1.amazonaws.com/s3fs-public/2020- 05/COVID-19-Tourism-Recovery-TAPackage_8%20May-2020.pdf

World Tourism Organization (May 2020) UNWTO World Tourism Barometer May 2020. Special Focus on the impact of COVID-19. Retrieved from: https://webunwto.s3.eu-west1.amazonaws.com/s3fs- public/2020-05/Barometer%20-%20May%20202020%20- %20Short.pdf
.

Internet

Rural Tourism as an example of sustainable tourism. Accessed 16 February 2023. Available at https://www.ceupe.com/blog/turismo-rural-turismo-sostenible.html

Sustainable Tourism and Rural Tourism. Accessed 17 February 2023. Available at https://www.ceupe.com/blog/turismo-sostenible-y-turismo- rural.html

Entrepreneurship as a factor in sustainable rural tourism development. Consulted

on 18 from February from 2023. Available en
http://scielo.sld.cu/scielo.php?script=sci_arttext&pid=s2306-
91552016000100006

Sustainable Rural Tourism as a Development Opportunity for Small
Communities in Developing Countries. Accessed 19 February 2023. Available at
https://www.uv.mx/blogs/uvi/2009/01/15/el-turismo-rural- sustainable-rural-
tourism-as-a-development-opportunity-for-small-communities-in-developing-
countries/

I want morebooks!

Buy your books fast and straightforward online - at one of world's fastest growing online book stores! Environmentally sound due to Print-on-Demand technologies.

Buy your books online at
www.morebooks.shop

Kaufen Sie Ihre Bücher schnell und unkompliziert online – auf einer der am schnellsten wachsenden Buchhandelsplattformen weltweit! Dank Print-On-Demand umwelt- und ressourcenschonend produziert.

Bücher schneller online kaufen
www.morebooks.shop

info@omniscriptum.com
www.omniscriptum.com

Printed by Books on Demand GmbH, Norderstedt / Germany